YOUR KNOWLEDGE HAS VALUE

- We will publish your bachelor's and master's thesis, essays and papers

- Your own eBook and book - sold worldwide in all relevant shops

- Earn money with each sale

Upload your text at www.GRIN.com
and publish for free

Bibliographic information published by the German National Library:

The German National Library lists this publication in the National Bibliography; detailed bibliographic data are available on the Internet at http://dnb.dnb.de .

This book is copyright material and must not be copied, reproduced, transferred, distributed, leased, licensed or publicly performed or used in any way except as specifically permitted in writing by the publishers, as allowed under the terms and conditions under which it was purchased or as strictly permitted by applicable copyright law. Any unauthorized distribution or use of this text may be a direct infringement of the author s and publisher s rights and those responsible may be liable in law accordingly.

Imprint:

Copyright © 2016 GRIN Verlag, Open Publishing GmbH
Print and binding: Books on Demand GmbH, Norderstedt Germany
ISBN: 9783668441088

This book at GRIN:

http://www.grin.com/en/e-book/357218/access-of-iodized-salt-for-households-of-rural-janamora-woreda

Souvenir Jean Jacques Bucyana, Aweke Kebede, Bamlak Lamilek

Access of iodized salt for households of rural Janamora Woreda

GRIN Publishing

GRIN - Your knowledge has value

Since its foundation in 1998, GRIN has specialized in publishing academic texts by students, college teachers and other academics as e-book and printed book. The website www.grin.com is an ideal platform for presenting term papers, final papers, scientific essays, dissertations and specialist books.

Visit us on the internet:

http://www.grin.com/

http://www.facebook.com/grincom

http://www.twitter.com/grin_com

Access of iodized salt for households of rural Janamora Woreda

Souvenir Jean Jacques Bucyana[a*] (MSc) ; Aweke Kebede[b] (PhD.) ; Bamlak Lamilek[c] (PhD.)

Authors' affiliations:

[a*]**Souvenir Jean Jacques Bucyana ;** Correspondant author, Addis Ababa University, College of development studies, Centre for food security studies, Addis Ababa, Ethiopia

[b]**Aweke Kebede,** Ministry of Health, Addis Ababa, Ethiopia

[c] **Bamlak Lamilek,** Addis Ababa University, College of development studies, Centre for rural development, Addis Ababa, Ethiopia.

Abstract

The study aimed at determining the quality of consumed salt among 320 households that have children of 6-24 months and live in rural areas of Janamora Woreda, Northwestern zone of Ethiopia. By examining the concentration of Iodine in the salt consumed by surveyed households, the papers revealed that access to adequate and nutritional food for a healthy and active life among the rural poor families is determined by just more than availability.

Key words: Iodized salt, micronutrient, Iodine deficiency disorder, food security, fortification.

1. Introduction...2
2. Methods and tools 3
3. Results and discussion 3
3.1 Household's demographic information..3
3.2 Consumption of iodized salt among Janamora Woreda households 4
4. Conclusion and recommendations 5
References..5

1. Introduction

An estimated 2 billion people worldwide have inadequate iodine intake, although international efforts have helped to reduce the most severe manifestations of iodine deficiency disorders (cretinism). Dietary deficiencies are associated with soils which are deficient in iodine, unless there are other sources in the diet such as marine products. Both industrialized and developing countries still exhibit deficiencies [2].

The elimination of Iodine Deficiency Disorders (IDD) consists the most important health and social goal. Iodine deficiency at critical stages during pregnancy and early childhood results in impaired development of the brain and consequently in impaired mental function.While a variety of methods exists for the correction of iodine deficiency, in practice the most commonly applied is universal salt iodization which consist in the addition of suitable amounts of potassium iodate to all salt for human and livestock consumption [1].

Fortification of salt with Iodine has been one of the longest standing micronutrient interventions which began following scientific proof of the effectiveness of iodine as a goiter prophylactic in the 1920's. Iodization of salt began in Switzerland in 1922 and in Michigan in the US in 1924. The Universal Salt Iodization initiative began a major extension of salt iodization to the developing world in 1990. This increased the proportion of the world's population with access to iodized salt from 20% prior to 1990, to about 70% currently [2]. In fact, a huge progress has been made, such that 34 countries have reached the universal salt iodization goal (USI is defined as when all salt for human and animal consumption is iodized to the internationally agreed recommended levels) (whereby 90 % of the population consume iodized salt), and at least 28 more have 70-89 % of the population covered [1]. However, iodine deficiency remains a public health problem in 47 countries [2].

Particularly, micronutrient malnutrition consist a severe public health problem in many countries of Africa including Ethiopia. The iodine results from the 2005 micronutrient survey, conducted by the Ethiopian Public Health Institute (EPHI) revealed that goiter prevalence among children was 40 % (28 % for palpable and 12 % for visible goiter) and 36 % among women aged 15-45 (EPHI, 2013). This survey also found that just 4.2 % of Ethiopian households had iodized salt, a marked deterioration from the previous decade [3].

Despite the fact that salt iodization has been recommended as the preferred and affordable strategy to control and eliminate Iodine Deficiency Disorders (IDD) [2] and intensive efforts have been made by the governments of IDD-affected countries to implement salt iodization programs over the last decade. However, there is a need to provide scientific evidence to demonstrate that adequate amounts of iodine are reaching the target population.This study therefore, aimed at assessing Iodine status among households that had children of 6-24 months of age living in rural areas of Janamora Woreda, Northwestern Zone of Ethiopia.

2. Methods and tools

A household survey was realized on 320 households that live in rural areas of Janamora Woreda and that have a child of 6-24 months of age. Five Kebeles namely Deresgie, Awuchara, Majji Sarabar, and Wayna were purposively selected due to their accessibility. Demographic as well as economic factors were investigated.Iodine status was assessed using Iodine test field kit to measure the concentration of Iodine in the salt.

3. Results and discussion

3.1 Household's demographic information

The survey was conducted on 320 households from rural lowland (<2500 m above sea level (a.s.l) and highland areas (>2500m.a.s.l and above) of Janamora Woreda representing a total population number of 1694.The results revealed that the level of literacy among all the household heads was at 40% (N=320). The greater number of illiterate household heads was from the highland areas with a literacy rate of 37.8%. The average family size in study area (5.3) was higher than national average of 4.9 for rural households [3] and 40 % (N=320) of the households were made of more than 5 members in their family (see Table 1). The minimum family size is 3 while the maximum was 10. The average of dependency ratio of the two agro-ecological areas was 120.5 (± 64.6).The minimum was 40 while the maximum value is 400. Dependency ratio relates the number of children (0-14 years old) and older persons (65 years or over) to the working-age population (15-64).It indicates the potential effects of changes in populations age structures for social and economic development, pointing out broad trends in social support needs.

3.2 Consumption of iodized salt among Janamora Woreda households

Among the highland households non-iodized salt (0 PPM) was prevalent at 6.5 % (n=185), the concentration of less than 15 PPM was 38.4 % (n1=185) for the highland and 42.3 % (n2=135) among lowland households, whereas the salt with Iodine of more than 15 PPM concentration was prevalent among highland households at a level of only 6.5 % (n=185) and at 57.7 % (n2=135) among lowland counterparts (See Figure 1). According to respondents, low purchasing power consists the main cause of not buying iodized salt as it is more expensive compared to the non-iodized. While the other reasons are lack of knowledge about iodized salt and lack of adequate supply. They reported that the two types of salts taste the same and that they do not perceive the importance to purchase the same products with different prices.

On the other hand low consumption of adequate and healthy food consist an evidence that access to adequate and healthy food is reverberated through entitlement. As key component of food security and nutrition, availability of food on the market is not a sufficient condition to access adequate food especially in the rural settings. Therefore, further interventions should go beyond prevailing marketing practices and involve the social aspect of people for example through engaging in educating people about the benefit of the products and through incentives to companies that provide such services. There is also a need to remove from the market the non-iodized salt or else to increase taxes on suppliers as a mean to discourage them.

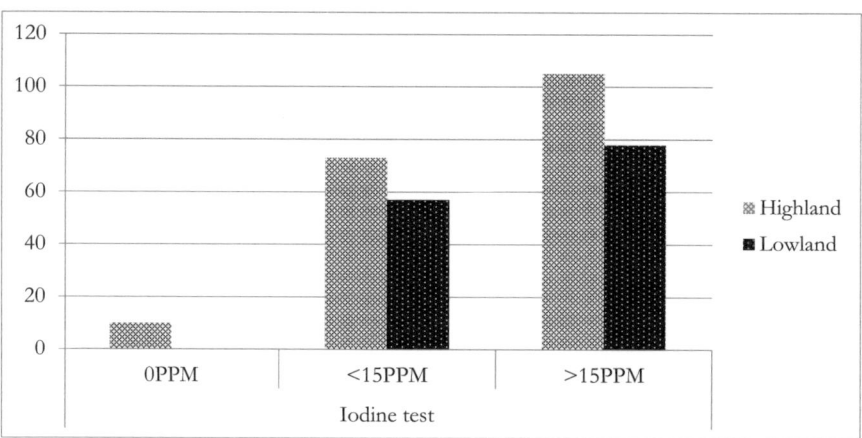

Figure 1 Consumption of iodized salt among sample households

4. Conclusion and recommendations

Iodized salt is a promising route towards IDD eradication; therefore salt fortification should be supported by both government of Ethiopia, NGOs and private institutions.Campaign to sensitize the consumption of iodized salt should be increased among rural communities.There is also a need for implementation of practical and effective surveillance to control micronutrient deficiency successfully especially in areas that are hard to reach.

References

[1] ICCID, UNICEF, WHO (2001) Assessment of Iodine Deficiency Disorders and monitoring their elimination: A guide for programme managers. The Netherlands.

[2] Sue Horton, Venkatesh Mannar, Annie Wesley (2008) Micronutrient fortification (Iron and Salt Iodization).Best practice paper: New advice from Copenhagen consensus centre. Denmark.

[3] EPHI (2013) Ethiopia National food consumption survey. Addis Ababa. Ethiopia.

[4] Federal Ministry of Health (2011) National guideline for family planning services in Ethiopia. Addis Ababa, Ethiopia.

Agro-ecological areas	HH's head sex (N=320)		HH's head age (N=320)		Literacy (N=320)		Total HH Members (N=320)	
	Female	Male	21-40 years	>41	Illiterate	Literate	<=5	> 5years
Highland (n1=185)	0.6% (2)	57.2% (183)	42.5% (136)	15.3 (49)	37.8% (121)	20% (64)	55.2% (102)	34% (63)
Lowland (n2=135)	0.3% (1)	41.9% (134)	35.3% (113)	6.9% (22)	22.5% (72)	19.7 (63)	57.8 (78)	42.3% (57)

Table 1. Households' demographic information

YOUR KNOWLEDGE HAS VALUE

- We will publish your bachelor's and master's thesis, essays and papers

- Your own eBook and book - sold worldwide in all relevant shops

- Earn money with each sale

Upload your text at www.GRIN.com and publish for free